Contents

Introduction 2
Testing Schedule 4
Test 1 5
Test 2 7
Test 3 9
Test 4 11
Test 5 13
Test 6 15
Test 7 17
Test 8 19
Test 9 21
Test 10 23
Test 11 25
Test 12 27
Test 13 29
Test 14 31
Test 15 33
Test 16 35
Test 17 37
Test 18 39
Test 19 41
Test 20 43
Test 21 45
Test 22 47
Test 23 49
Test Answer Forms 51
Test Analysis Form 57
Test Solutions 59

Introduction

The Saxon Homeschool Testing Book for Algebra 1 contains Tests, a Testing Schedule, Test Answer Forms, a Test Analysis Form, and Test Solutions. Descriptions of these components are provided below.

About the Tests

The tests are available after every five lessons, beginning after lesson 10. The tests are designed to provide students with sufficient time to learn and practice each concept before they are assessed. The test design allows students to display the skills they have developed, and it fosters confidence that will benefit students when they encounter comprehensive standardized tests.

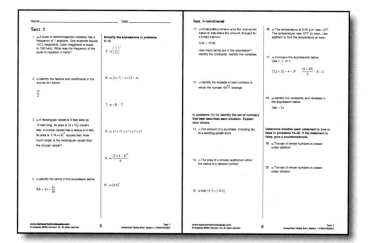
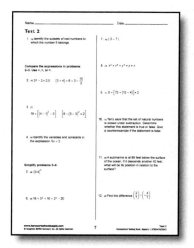

Testing Schedule

Administering the tests according to the schedule is essential. Each test is written to follow a specific five-lesson interval in the textbook. Following the schedule allows students sufficient practice on new topics before they are assessed on those topics.

Tests should be given after every fifth lesson, beginning after Lesson 10. The testing schedule is explained in greater detail on page 4 of this book.

Optional Test Solution Answer Forms are included in this book. Each form provides a structure for students to show their work.

About the Test Solution Answer Forms

This book contains three kinds of answer forms for the tests that you might find useful. These answer forms provide sufficient space for students to record their work on tests.

Answer Form A: Test Solutions

This is a double-sided master with a grid background and partitions for recording the solutions to twenty problems.

Answer Form B: Test Solutions

This is a double-sided master with a plain, white background and partitions for recording the solutions to twenty problems.

Answer Form C: Test Solutions

This is a single-sided master with partitions for recording the solutions to twenty problems and a separate answer column on the right-hand side.

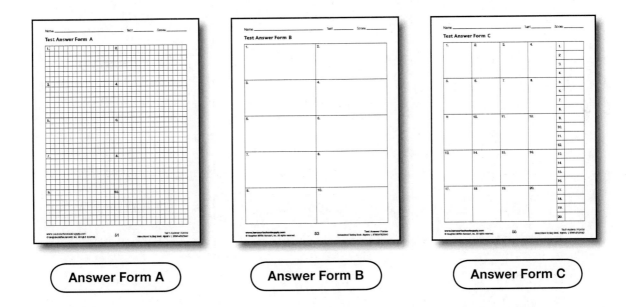

Answer Form A Answer Form B Answer Form C

Test Solutions

The Test Solutions are designed to be representative of students' work. Please keep in mind that problems may have more than one correct solution. We have attempted to stay as close as possible to the methods and procedures outlined in the textbook.

Testing Schedule

Test to be administered	Covers material through	Give after teaching
Test 1	Lesson 5	Lesson 10
Test 2	Lesson 10	Lesson 15
Test 3	Lesson 15	Lesson 20
Test 4	Lesson 20	Lesson 25
Test 5	Lesson 25	Lesson 30
Test 6	Lesson 30	Lesson 35
Test 7	Lesson 35	Lesson 40
Test 8	Lesson 40	Lesson 45
Test 9	Lesson 45	Lesson 50
Test 10	Lesson 50	Lesson 55
Test 11	Lesson 55	Lesson 60
Test 12	Lesson 60	Lesson 65
Test 13	Lesson 65	Lesson 70
Test 14	Lesson 70	Lesson 75
Test 15	Lesson 75	Lesson 80
Test 16	Lesson 80	Lesson 85
Test 17	Lesson 85	Lesson 90
Test 18	Lesson 90	Lesson 95
Test 19	Lesson 95	Lesson 100
Test 20	Lesson 100	Lesson 105
Test 21	Lesson 105	Lesson 110
Test 22	Lesson 110	Lesson 115
Test 23	Lesson 115	Lesson 120

Test 1

1. (3) A burst of electromagnetic radiation has a frequency of 1 exahertz. One exahertz equals 1012 megahertz. Each megahertz is equal to 106 hertz. What was the frequency of the burst of radiation in hertz?

2. (2) Identify the factors and coefficients in the expression below.

    $$\frac{m}{2}$$

3. (4) A rectangular carpet is 9 feet wide by 12 feet long. Its area is $(9 \cdot 12)$ square feet. A circular carpet has a radius of 4 feet. Its area is $3.14 \cdot (4)^2$ square feet. How much larger is the rectangular carpet than the circular carpet?

4. (2) Identify the terms in the expression below.

    $$8jk + 2x + \frac{4j}{3k}$$

Simplify the expressions in problems 5–10.

5. (3) $\left(\dfrac{1}{11}\right)^2$

6. (4) $(2 \cdot 7) + 4 \cdot (3 + 4)$

7. (5) $-|9 - 7|$

8. (3) $x^3 \cdot x^2 \cdot y^4 \cdot y^2 \cdot y^3$

9. (4) $\dfrac{(2 \cdot 4 - 5)^3}{9}$

10. (3) $(0.5)^3$

Test 1—continued

11. (2) A babysitting service uses the expression below to determine the amount charged for a single session.

    $9.0h + 15.95$

    How many terms are in the expression? Identify the constants. Identify the variables.

12. (1) Identify the subsets of real numbers to which the number $6\sqrt{11}$ belongs.

In problems 13–14, identify the set of numbers that best describes each situation. Explain your choice.

13. (1) The amount of a purchase, including tax, at a sporting goods store

14. (1) The area of a circular auditorium when the radius is a rational number

15. (5) Add $(4.1) + (-6.3)$.

16. (5) The temperature at 6:00 a.m. was −2°F. The temperature rose 10°F by noon. Use addition to find the temperature at noon.

17. (4) Compare the expressions below. Use <, >, or =.

    $(3.2 + 2) \div 4 + 3^2 \;\bigcirc\; \dfrac{(9+23)}{2} - 8 \div 2$

18. (2) Identify the constants and variables in the expression below.

    $3ab + 2x$

Determine whether each statement is true or false in problems 19–20. If the statement is false, give a counterexample.

19. (5) The set of whole numbers is closed under addition.

20. (1) The set of whole numbers is closed under division.

Test 2

1. (1) Identify the subsets of real numbers to which the number 6 belongs.

Compare the expressions in problems 2–3. Use <, >, or =.

2. (4) $3^2 - 2 \cdot 2.5 \bigcirc (3+4) + 6 \div 3 - \dfrac{10}{2}$

3. (7) $18 \div \left[(4-1)^2 - 3\right] \bigcirc \left[8 - (5-3)^2 \cdot 2\right]$

4. (2) Identify the variables and constants in the expression $5x + 2$.

Simplify problems 5–9.

5. (3) $(0.4)^3$

6. (4) $18 + 3^3 + 16 \div 2^3 - 20$

7. (5) $|2 - 7|$

8. (3) $x^2 \cdot x^6 \cdot y^5 \cdot y \cdot x$

9. (7) $6 + \left[72 \div (12 - 4)\right] \cdot 2$

10. (1) Terry says that the set of natural numbers is closed under subtraction. Determine whether this statement is true or false. Give a counterexample if the statement is false.

11. (6) A submarine is at 89 feet below the surface of the ocean. If it descends another 42 feet, what will be its position in relation to the surface?

12. (6) Find the difference $\left(\dfrac{2}{7}\right) - \left(-\dfrac{4}{7}\right)$.

Test 2–continued

13. (8) Charlie walks at a rate of 7040 yards per hour. How fast does Charlie walk in miles per hour?

14. (2) Identify the terms in the expression $\dfrac{16a}{(2+b)} - c + 2ac - \dfrac{b}{4}$.

Evaluate each expression in problems 15–16 for the given values.

15. (9) $2n + 3n + mn$ for $m = 4$ and $n = 3$

16. (9) $2(x - y)^3 - x^2$ for $x = 5$ and $y = 2$

17. (10) Order the numbers below from least to greatest.

$-0.5, \dfrac{1}{2}, 0.7, -\dfrac{3}{4}$

18. (10) The table shows rainfall for 4 different locations in 2006 and 2007. Which location had the greatest increase in rainfall from 2006 to 2007?

Rainfall (inches)		
	2006	2007
Location A	45.6	47.2
Location B	39.9	40.7
Location C	38.6	39.2
Location D	49.9	51.3

19. (8) The volume of a barrel of sand is 18 cubic feet. How many barrels of sand will be needed to fill a container with a volume of 2 cubic yards?

20. (5) One day in January the temperature at 8 a.m. was −5°F. By noon, the temperature had risen 15 degrees. What was the temperature at noon?

Test 3

1. *(12)* Identify the property illustrated.

   $(5 \cdot 27) \cdot 2 = (27 \cdot 5) \cdot 2$

2. *(9)* Compare the expressions when $x = 3$ and $y = 2$. Use <, >, or =.

   $x^2 + y^3 \bigcirc x^2 y^2$

3. *(6)* Find the difference $-1.1 - (-2.3)$.

4. *(Inv. 1)* Lily spun a game spinner and recorded the results in the table below.

Outcome	Frequency
1	7
2	5
3	8

   What is the probability of landing on 1? on 2? on 3? Express each probability as a fraction in simplest form and a percent.

Simplify problems 5–10.

5. *(11)* $(-7)(-0.2)$

6. *(11)* -3^2

7. *(15)* $4(8 - 6)$

8. *(15)* $-2(3 + 5)$

9. *(3)* $\left(\dfrac{1}{7}\right)^2$

10. *(10)* $-\dfrac{3}{13} - \left(-\dfrac{7}{13}\right) + \dfrac{2}{13} - \dfrac{1}{13}$

Test 3–continued

11. *(14)* In a bag, there are 10 tiles numbered as follows: 1, 2, 2, 2, 2, 2, 3, 3, 4, and 4. A tile is randomly chosen from the bag. What is the probability of drawing a tile with a number greater than 2?

12. *(13)* Is 48 a perfect square? Explain.

13. *(12)* Tell whether the statement $c + 1 = c$ is true or false. Justify your answer using properties. Assume c is a real number.

14. *(Inv. 1)* Describe the following event as impossible, unlikely, as likely as not, likely, or certain: Jane rolls an even number on a number cube labeled 1–6.

15. *(1)* Identify the subset of real numbers to which $2\sqrt{5}$ belongs.

16. *(5)* Find the sum $(-12) + (-7)$.

17. *(4)* A rectangular mirror is 11.5 inches by 10 inches. Its area is $(11.5 \cdot 10)$ square inches. A circular mirror has a radius of 5 inches. Its area is approximately $3.14 \cdot 5^2$ square inches. What is the combined area of the two mirrors?

18. *(8)* Kelly rows a boat at the rate of 7040 yards per hour. How fast did Kelly row in miles per hour?

19. *(14)* A number cube labeled 1–6 is rolled. List the outcomes for the event that a number greater than 5 is rolled.

20. *(13)* Estimate the value of $\sqrt{17}$ to the nearest integer.

Name _____   Date _____

Test 4

1. (17) Write "c increased by 5" as an algebraic expression.

Evaluate each expression in problems 2–3 for the given values of the variables.

2. (16) $(-x - y) - (x + y)$ for $x = 3$ and $y = 2$.

3. (16) $(ab) + (abc)$ for $a = -2$, $b = 2$, and $c = -3$.

4. (20) Graph the ordered pair $(-3, 1)$ on the coordinate plane.

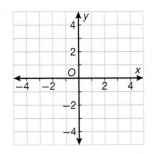

Simplify problems 5–10.

5. (18) $-3x - (-2x) + 4x$

6. (18) $6n^2 - 4m^2 - 2n^2 + 9m^2$

7. (15) $-8(y - 7)$

8. (11) $(-3)(-8)$

9. (10) $-2.86 + (-3.24) - 4.71 + 6.62$

10. (7) $7 + 3\left[(9 - 2)^2 - 5\right]$

11. (8) A horse gallops at a speed of 45,760 yards per hour. How fast does the horse gallop in miles per hour? (Hint: A mile is equal to 1760 yards.)

Test 4—continued

12. (19) Tell whether $x = 4$ is a solution for the equation $x + 4 = 9$.

13. (19) Solve $a - 8 = 7$.

14. (6) Determine whether the set of integers is closed under subtraction. If the statement is false, give a counterexample.

15. (5) Find the sum $(-8) + (-10)$.

16. (13) Compare the expressions below. Use <, >, or =.

$$\sqrt{16} + \sqrt{49} \bigcirc \sqrt{25} + \sqrt{36}$$

17. (14) There are 3 blue, 4 red, and 3 yellow marbles in a bag. What is the probability of randomly choosing a red marble?

18. (17) Use words to write the algebraic expression $m - 9$ in two different ways.

19. (11) If a baseball player has 5 hits in 20 at-bats, what is the probability he will get a hit in the next at-bat? Express your answer as a decimal to the thousandths place.

20. (20) Complete the table for the equation $y = 2x + 4$.

x	y
–2	
0	
2	
$\frac{1}{2}$	

Test 5

1. (8) A train travels at a rate of 123,200 yards per hour. How fast does the train travel in miles per hour? (Hint: A mile is equal to 1760 yards.)

2. (2) Identify the terms in the expression $7ab + 9c + \dfrac{12a}{5b}$.

3. (22) The stem-and-leaf plot shows the ages of people who attended a party. Find the median age of people at the party.

Stem	Leaf
1	9
2	8 9 9
3	0 1 3 4 6
4	0 3 4 4 6
5	0 3 4 5

4. (13) The area of a square rug is 225 square feet. What is the side length of the rug?

Simplify problems 5–11.

5. (3) $x^6 \cdot x^3 \cdot y^2 \cdot y^4 \cdot x^2$

6. (4) $4^3 + 9 \div 3 - 2 \cdot (3)^2$

7. (5) $-|10 - 2|$

8. (7) $6 + 2\left[(5-3)^3 + 4\right]$

9. (11) $4.5 \div (-9)$

10. (15) $-xy(x^2 - yz^2)$

11. (18) $4a^4 + 2b^3 + 2a^4 + 3b^3$

Test 5—continued

12. (25) Walt travels at an average of 45 miles per hour on a 670-mile trip. Write a rule in function notation to find the number of miles he has left to travel at the end of any given hour. Let h represent the number of hours spent traveling.

13. (14) There are 5 red marbles, 3 blue marbles, and 2 green marbles in a bag. If a marble is chosen randomly, what is the probability that it is not green?

14. (20) Complete the table for the equation $y = 5x + 2$.

x	y
−2	
0	
2	
$\frac{3}{5}$	

Solve problems 15–18.

15. (21) $-6 = \dfrac{1}{3}x$

16. (23) $\dfrac{1}{3}b - \dfrac{1}{2} = \dfrac{3}{5}$

17. (24) $0.3r + 0.2 = 1.7$

18. (19) $-9 = y + 18$

For problems 19–20, evaluate each expression for the given values of the variables.

19. (16) $\dfrac{x(3yz)}{xz}$ for $x = 4$, $y = 3$, and $z = -2$.

20. (9) $3(x - y)^2 + 5y^3$ for $x = 6$ and $y = 2$.

Test 6

1. (8) The area of a garden is 30 square yards. If Joe can weed about 5 square feet per minute, how many minutes will it take him to weed the garden?

2. (22) A total of 700 tickets for the school play were sold. The circle graph shows the percentage of tickets sold by students in each class. Find the number of tickets sold by students in each class.

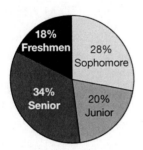

Tickets Sold by Class

3. (10) The line graph shows the number of customers at a restaurant for 6 days. Explain why the graph may be misleading.

4. (25) Determine the domain and the range of the relation $\{(2, 8), (9, 7), (2, 8), (6, 1), (2, 4)\}$.

Simplify problems 5–9.

5. (3) $\left(\dfrac{1}{3}\right)^3$

6. (4) $(6 \cdot 2) + 3 \cdot (4 + 1)$

7. (10) $\dfrac{5}{9} - \left(-\dfrac{1}{9}\right) + \dfrac{2}{9} - \dfrac{4}{9}$

8. (5) $\left| 1 - \dfrac{2}{3} \right|$

9. (18) $-3a - (-6a) + 2a$

Test 6–continued

10. (29) Solve $3x + 5y = 2$ for y.

11. (17) Write "6 more than the quotient of 4 divided by p" as an algebraic expression.

12. (14) A number cube labeled 1–6 is rolled. List the outcomes for the event that a number greater than 4 is rolled.

13. (1) Identify the subsets of real numbers to which the number $\frac{3}{4}$ belongs.

14. (30) Use the graph to identify the domain and range of the function.

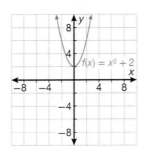

Solve problems 15–19.

15. (26) $a + 4(3a + 7) = 80$

16. (28) $6b = b - 5$

17. (21) $3y = -15$

18. (23) $8a + 3 = 27$

19. (24) $0.006s + 0.02 = 0.2$

20. (16) Evaluate $-a\left[b(a-b)\right]$ for $a = 5$ and $b = 4$.

Test 7

1. (31) Which is the better buy: 3 lbs of tomatoes for $2.85 or 5 lbs for $4.65?

2. (23) Mary wants to buy a bike for $290. She already has $110. If her job pays $15 per hour, how many hours will she have to work to earn the rest of the money?

3. (33) The spinner below is spun twice. Make a tree diagram to show all possible outcomes. What is the probability of spinning 2 both times?

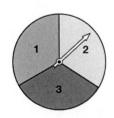

4. (Inv. 2) Describe the shape of the graph that represents the following situation: A rocket is launched up into the air, then falls back to the earth. The graph relates time to the rocket's distance from the earth.

Simplify problems 5–6.

5. (32) $\dfrac{a^{-6}}{b^5}$

6. (10) $2.38 + (-1.92) - (-6.22) + 3.47$

7. (8) Mr. Ramirez is buying square stones to cover his patio. The patio measures 7.5 feet by 9 feet. Each stone is 6 inches on a side. If Mr. Ramirez buys 275 stones, are there enough stones to cover the patio?

8. (12) Identify the property illustrated: $(4 \cdot 11) \cdot 9 = (11 \cdot 4) \cdot 9$.

9. (17) Write $\dfrac{3}{p}$ in words in two different ways.

10. (30) Graph the equation $y = x + 1$ using a table. Decide whether the graph represents a function and whether it is linear or nonlinear.

Test 7–continued

11. (34) A theater has group prices for tickets. The first ticket will cost $8 while each additional ticket will cost $5. Write a rule to model the situation. Then use the rule to find the total cost of 13 tickets. Let n represent the number of tickets and c represent the total cost.

12. (20) Graph the ordered pair (–2, 3) on a coordinate plane.

13. (35) Find the x- and y-intercepts for $2x + 3y = 18$.

14. (24) 0.27 of 45 is what number?

15. (21) Solve $\frac{2}{3}x = 8$.

Solve problems 16 and 17. Justify each step.

16. (28) $6c - 3 - 2c = 4c + 4$

17. (26) $5d + 2(2d + 3) = 42$

18. (13) Estimate the value $\sqrt{77}$ to the nearest integer.

19. (16) Evaluate $p(3p - 2q) - pq$ for $p = \frac{1}{2}$ and $q = \frac{1}{4}$.

20. (29) Solve for s: $4s - t - 5 = 2s + 3t$.

Test 8

1. (38) Find the prime factorization of 130.

2. (Inv. 3) A school conducts a survey about favorite sports. The coach asks members of the basketball team to name their favorite sport. Give a reason why the sampling method may be biased.

3. (27) A bookstore conducted a survey of the reading preferences of its customers. The bar graph shows the results. Explain why the graph may be misleading.

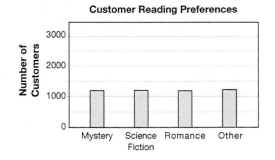

4. (37) Find the product $(2.7 \times 10^3)(2.4 \times 10^4)$. Write the answer in scientific notation.

Simplify problems 5–8.

5. (40) $(6x^2y^3)^2$

6. (39) $\dfrac{x^{-2}}{m}\left(\dfrac{y}{x^{-4}n^{-3}} + 3x^{-2}n^{-3}\right)$

7. (18) $2a^2 + 4a^2 + 3ab^2 - 3a^2 - 2ab^2$

8. (11) -4^2

9. (29) The formula $F = \dfrac{9}{5}C + 32$ expresses Fahrenheit temperature in terms of Celsius temperature. Find the Celsius temperature when the Fahrenheit temperature is 59°.

10. (36) A traffic light casts a shadow 23.4 feet long. A girl who is 4 feet tall casts a shadow 5.2 feet long. The triangle drawn with the traffic light and its shadow is similar to the triangle drawn with the girl and her shadow. How tall is the traffic light?

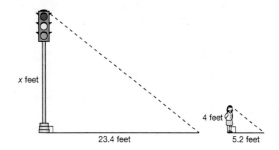

Test 8–continued

11. (33) A bag contains 3 green marbles and 7 yellow marbles. Find the probability of drawing a green marble, keeping it, and drawing another green marble. Write the probability as a fraction in simplest form.

12. (8) A traveler from Europe wants to convert the distance 700 miles to kilometers. If 1 mile equals 1.61 kilometers, what is the distance between the two cities in kilometers?

13. (34) Determine whether the sequence below is an arithmetic sequence.

    1, 12, 23, 34, . . .

    If yes, find the common difference and the next two terms.

14. (6) Find the difference $(-2.2) - (-6.7)$.

Solve problems 15–17.

15. (28) $-3(3y - 7) = 21 - 9y$

16. (26) $5b - 6 - 7b + 2 = 2$

17. (24) $-0.3c + 0.5 = 2.6$

18. (10) Order the numbers from least to greatest.

    $-\dfrac{1}{2}, -2, -0.2, 0.2$

19. (9) Evaluate $6c - 4d + 3cd$ for $c = 5$ and $d = 3$.

20. (17) Jason scores 3 points for each question answered correctly on an exam. Write an algebraic expression for the total points he will score if he answers n questions correctly.

Test 9

1. (38) Find the GCF of $12x^3y^4z^2 + 8x^2y^5z^3$.

2. (40) A square patio has a side length of $11n$ feet. What is the area of the patio?

3. (44) Determine the slope of the given line.

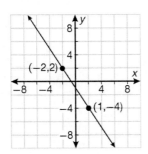

4. (45) Translate the sentence below into an inequality.

   The product of a number and 3 is greater than 5.

Simplify problems 5–8.

5. (39) $\dfrac{m^2}{n}\left(\dfrac{m^2}{n^2} + \dfrac{2n^3}{x}\right)$

6. (32) $\dfrac{1}{y^{-3}}$

7. (18) $6xy^2 + 5x^2y - 3yx^2 - 3y^2x$

8. (12) $(37) \cdot 9x \cdot \left(\dfrac{1}{37}\right)$

9. (41) A hiker walks at a constant speed, as shown in the table below. What is the rate of change?

Time (hours)	2	4	6	8
Distance (miles)	6	12	18	24

Test 9—continued

10. (42) Concert tickets sell for $12.50 each. On Friday night, 375 tickets are sold. If 20% of the money earned from ticket sales is donated to charity, how much is donated to charity?

11. (36) The figures below are similar. Find x.

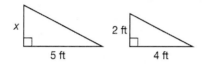

12. (43) Determine the values for which the rational expression $\frac{3a+2}{a+5}$ is undefined.

13. (27) A town created the graph below to show the average cost of property taxes each year. Explain why the graph may be misleading.

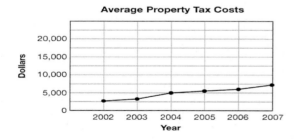

14. (Inv. 4) Provide a counterexample for the following statement: If a student likes football, then he is on the football team.

Solve problems 15–19.

15. (31) $\frac{2}{3} = \frac{y-7}{9}$

16. (28) $-3n = 2n + 5$

17. (26) $3z - (2 - z) - 5 = 9$

18. (23) $3 = -6a - 1$

19. (19) $\frac{3}{4} - a = \frac{1}{10}$

20. (35) At a video store, movie rentals are $4 and game rentals are $6. Let x be the number of movies rented. Let y be the number of games rented. The equation $4x + 6y = 240$ shows that the store made $240 renting movies and games. Find the intercepts and explain what each means.

Test 10

1. (50) Graph the inequality $x \leq -0.5$.

2. (33) A bag contains 3 red marbles and 3 blue marbles. Find the probability of drawing a red marble, keeping it, and drawing another red marble. Write the probability as a fraction in simplest form.

3. (48) Find the mean, median, and mode of the values in the data set below, rounded to the nearest whole.

    42, 44, 38, 44, 57, 39, 44, 39, 41

4. (2) Identify the factors and coefficients in the expression $\dfrac{x}{7}$.

Simplify problems 5–7.

5. (46) $\sqrt{\dfrac{16}{81}}$

6. (39) $\dfrac{xy}{a}\left(\dfrac{xyz}{b} + 3ax - b^2\right)$

7. (40) $\left(2a^2b\right)^4$

8. (38) Factor $7x^3y^4 - 7x^2y^5$ completely.

9. (41) Find the slope of the line below.

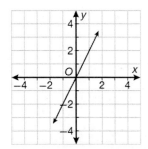

10. (44) Determine the slope of the line that contains the points (1, 5) and (9, 3).

Test 10—continued

11. (49) The equation $y = -2x + 1$ is in slope-intercept form. Graph $y = -2x + 1$ on a coordinate grid.

12. (47) A jewelry store marks up the price of bracelets they purchase at $30.00 each by 75%. What is the markup and new price of each bracelet?

13. (43) Determine the value for which the expression $\dfrac{(a+2)(a-3)}{5a+15}$ is undefined.

14. (37) Write 0.0056 in scientific notation.

15. (34) Write the first 4 terms of an arithmetic sequence where $a_1 = -4$ and the common difference $d = 5$.

16. (28) Health Club A charges a $100.00 membership fee plus $25.00 per month. Health Club B charges a $200.00 membership fee plus $20.00 per month. For what number of months are the health club costs the same?

17. (31) The ratio of men to women in a cooking class is 7 : 5. In all, the class has 24 students. How many men and how many women are in the class?

18. (Inv. 3) Directors of a library want to know what people think of their library. They survey every fifth person who enters and checks out a book. What is a possible bias for this survey?

19. (45) Write the inequality $6 \leq 9b$ in words.

20. (Inv. 5) Identify Statement 2 as the *converse*, *inverse*, *contrapositive*, or *contradiction* of Statement 1. Then indicate the truth value of each statement.

    Statement 1: If a figure is a square, then it has four sides.

    Statement 2: If a figure is a square, then it does not have four sides.

Test 11

1. (45) Kendra is buying gifts for 5 friends. She can spend $100 at most. She has already spent $28 on one gift. If she wants to spend the same amount on each of the other 4 gifts, what can she spend at most on each gift?

2. (54) The box-and-whisker plot shows heights in inches of students in a math class. Use the interquartile range to identify outliers.

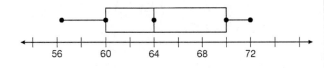

3. (51) Simplify $\dfrac{2x^2}{8x}$ if possible. Identify any excluded values.

4. (48) The following data show weights in pounds of some dogs. Identify any outliers. What is the effect of any outliers on measures of central tendency?

   35, 45, 46, 39, 31, 40, 3, 36

Simplify problems 5–8.

5. (40) $(2x^3y)^3$

6. (18) $2x^3 + 3y^2 + 4x^3y^2 - 3x^3 - 2x^3y^2$

7. (39) $\dfrac{m^2}{n}\left(\dfrac{n^4}{m} + \dfrac{4n^2}{m^2}\right)$

8. (32) $\dfrac{1}{c^{-2}}$

9. (53) Write the polynomial $4x^2 + 2x^3$ in standard form. Then write the leading coefficient.

10. (44) Determine the slope of the line that contains the points (−4, 6) and (1, 5).

Test 11–continued

11. (55) Tell whether the ordered pair (2, 4) is a solution of the system below.

    $x - 2y = -6$
    $3x + 2y = 12$

12. (49) Write the equation of the graphed line in slope-intercept form.

    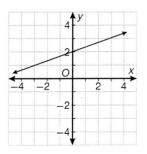

13. (46) A block in the shape of a cube has a volume of 8 cubic inches. What is the side length of the block?

Solve problems 14–16.

14. (31) $\dfrac{2}{7} = \dfrac{z-5}{35}$

15. (28) $-12 - 18c = -6(3c + 2)$

16. (23) $4 = -2t - 8$

17. (47) A frame store purchases frames for $12.00 each and then marks up the price of each by 45%. What is the markup and new price of each frame?

18. (41) Rashid reads a book at a constant speed, as shown in the table below. What is the rate of change?

Time (minutes)	5	10	15	20
Pages	10	20	30	40

19. (52) Write in point-slope form the equation of a line that has a slope of 2 and passes through point (3, 1).

20. (34) Determine whether the sequence below is an arithmetic sequence.

    6, 1, –4, –9, . . .

    If yes, find the common difference and the next two terms.

Test 12

1. (60) A photograph is in the shape of a rectangle. It has a length of $(x+3)$ inches and a width of $(x-3)$ inches. What is the area of the photograph?

2. (Inv. 6) Graph $f(x) = x$ and $f(x) = x + 1$ on the same coordinate grid. Compare the graph of $f(x) = x + 1$ to the parent graph. Use the y-intercept in your comparison.

3. (53) Find the degree of the monomial $7xy^2z^2$.

4. (58) Multiply $-3ab(2a^2 + 4b + 3c)$.

Simplify problems 5–8.

5. (51) $\dfrac{3x}{d^3} - \dfrac{4x}{d^3}$

6. (46) $\sqrt[3]{125}$

7. (43) $\dfrac{3y^2 - 12y}{9y - 36}$

8. (39) $\dfrac{a^{-2}}{b}\left(\dfrac{ab}{c^{-3}} + \dfrac{cb^3}{a^{-3}}\right)$

9. (50) Determine which of the values below are part of the solution set of the inequality $5 - 2x < 9$.

 $\{-2, -1, 0, 1, 2\}$

10. (57) Find the LCM of 7, 8, and 10.

11. (59) What is the solution to the system of equations?

 $y = 4x - 1$
 $y = 2x + 1$

Test 12–continued

12. (44) Determine the slope of the line that contains the points shown in the table.

x	y
−4	15
4	9
8	6

13. (47) A shoe store is having a sale of 25% off all boots. What is the discount and new price of a pair of boots that originally cost $95?

14. (45) Translate the sentence below into an inequality.

The product of $\frac{5}{8}$ and a number is less than or equal to 15.

15. (41) Find the slope of the horizontal line.

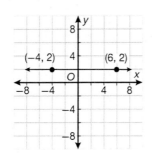

16. (42) What number is 110% of 90?

17. (56) An orchard sells 4 pounds of apples for $1. Graph the relationship and use the graph to estimate the cost of 5 pounds of apples.

18. (48) The table shows the math test scores for two classes at a school. Does Class A or Class B have a greater range of scores for the math test?

Scores on Math Test

Class A	91, 67, 98, 77, 58, 65, 88, 85, 81, 79, 63, 99, 82, 76, 100
Class B	87, 56, 87, 43, 91, 87, 58, 80, 86, 78, 74, 82, 67, 78, 65

19. (49) Determine the slope and y-intercept of the equation $3x + 4y = 12$.

20. (52) Denise is selling roses. She sold 3 for $5 and 5 for $7. What will Denise charge for 12 roses if she keeps selling roses at the same rate?

Test 13

1. (47) A clothing store is having a sale of 30% off all dresses. What is the discount and new price of a dress that originally cost $90?

2. (61) Simplify $\sqrt{18}$ using perfect squares.

Find the product for problems 3–4.

3. (60) $(y + 7)(y - 7)$

4. (58) $-2xy(-4y^2 + 3x + 5z)$

5. (18) Simplify the expression

   $7m^2n^2 - 5m^2n + 3m^2n - 4m^2n^2$.

6. (49) The equation $y = 3x - 3$ is in slope-intercept form. Graph $y = 3x - 3$ on a coordinate grid.

7. (46) A gift box in the shape of a cube has a volume of 216 cubic inches. What is the side length of the gift box?

8. (45) Translate the sentence below into an inequality.

   12,306 is less than or equal to the product of 3000 and a number.

9. (36) The figures below are similar. Find $m\angle D$.

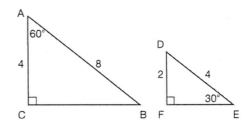

10. (65) Write an equation in slope-intercept form for the line that passes through (−3, 1) and is parallel to $y = 2x - 4$.

11. (62) Some children are at a summer camp. The heights of the children in inches are shown below. Create a stem-and-leaf plot of the data.

    50, 48, 36, 38, 45, 62, 52, 54, 48, 62, 48
    45, 47, 39, 44, 62, 65, 56, 53, 52, 42, 60,
    39, 54, 59, 45, 60, 49, 60, 61

Test 13—continued

12. (59) Solve the systems of equations by substitution.

    $2x + y = 0$

    $3x + 4y = 10$

13. (63) Solve the system of equations by elimination.

    $-3x + 6y = 15$

    $3x + 2y = 17$

14. (52) Write in point-slope form the equation of a line that has a slope of 3 and passes through point (3, −3).

15. (43) Determine the values for which the rational expression $\dfrac{7x - 1}{x - 2}$ is undefined.

16. (57) Gloria goes to the gym every 6 days. Liz goes to the gym every 4 days. Nate goes to the gym every 8 days. If all three go to the gym on the same day, in how many days will they all go to the gym on the same day again?

17. (64) If y varies inversely as x and $y = 8$ when $x = 7$, find x when $y = 4$.

18. (48) The following data show scores for a football team. Identify any outliers. What is the effect of any outliers on measures of central tendency?

    17, 6, 12, 14, 9, 67, 13, 17, 27

19. (38) Factor $-4x^2y^4 + -4x^3y^3$ completely.

20. (56) Write an equation for a direct variation that includes the point (4, 16).

Test 14

1. (61) A piece of fabric is in the shape of a square. The area measures 28 square inches. Find the length of one side of the piece of fabric.

2. (31) A car travels at 30 miles per hour. What is its rate in miles per minute?

Simplify problems 3–6.

3. (69) $2\sqrt{xy} - 9\sqrt{xy}$

4. (7) $3 - 4\left[(8-2)^2 + 4\right]$

5. (32) $\dfrac{a^{-2}}{b^4}$

6. (58) $(2x-2)(-3x+3)$

7. (36) A radio tower casts a shadow 33 feet long. A boy who is 5 feet tall casts a shadow 5.5 feet long. The triangle drawn with the radio tower and its shadow is similar to the triangle drawn with the boy and his shadow. How tall is the radio tower?

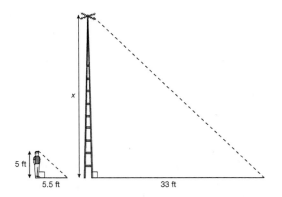

8. (65) Determine whether the lines are parallel.

$y = \dfrac{2}{5}x + 6$

$-2x + 5y = 30$

Solve the inequalities in problems 9–10 and graph them on a number line.

9. (66) $x - 3 < -1$

10. (70) $-\dfrac{x}{3} < 2$

Test 14–continued

11. (64) Nick is typing a research paper on the computer. His typing speed is inversely related to his typing time. If he is typing at 45 words per minute, it will take him 4 hours to type his entire paper. How long will it take him if he types 30 words per minute?

12. (40) A square photograph has a side length of $5n$ centimeters. What is the area of the photograph?

13. (63) Solve the system of equations by elimination.

    $2x + 8y = 12$

    $6y = 4x - 2$

14. (41) A boat travels at a constant speed as shown in the table below. What is the rate of change?

Time (hours)	2	4	6	8
Distance (miles)	70	140	210	280

15. (44) Determine the slope of the line that contains the points (–1, 4) and (2, –5).

16. (47) A shoe store marks down the price of a $35.00 pair of sandals by 25%. What is the discount and the new price of the pair of sandals?

17. (54) Make a box-and-whisker plot to display data of student scores on a math test.

    75, 62, 65, 90, 77, 76, 89, 64, 72, 68, 95, 77, 85, 70, 64

    Half of the students scored between which scores?

18. (68) What is the probability of rolling either two identical numbers or rolling a sum of 6 using two number cubes, each labeled 1–6?

19. (Inv. 7) Identify the constant of variation if $y = 6$ when $x = 3$, given that y varies directly with x. Then write the equation of variation.

20. (67) Solve the system of equations below.

    $y = -\dfrac{3}{4}x - 3$

    $3x + 4y = 12$

Test 15

1. (63) Sheila bought 15 pieces of fruit, some peaches and some plums. Peaches cost $0.89 each and plums cost $0.39 each. If Sheila spent $8.85 on fruit, how many plums did she buy?

2. (71) Use the data in the table to make a scatter plot. Draw a trend line on the scatter plot. Find an equation for the trend line.

x	1	2	3	4	5	6
y	1	3	8	11	13	16

Simplify problems 3–6.

3. (61) $\sqrt{81c^2d^3}$

4. (40) $(3x^3y^2)^2$

5. (39) $\dfrac{m^2}{n^2}\left(\dfrac{n^3}{m^2} + \dfrac{5n^3}{m}\right)$

6. (46) $\sqrt{\dfrac{4}{121}}$

7. (73) Write a compound inequality that represents all real numbers that are greater than 1 and less than 6. Graph the solution.

8. (65) Determine whether the lines passing through the following points are perpendicular.

   Line 1: (−1, 5) and (−3, 2)

   Line 2: (−1, 3) and (−4, 5)

   Explain your reasoning.

Factor the trinomials in problems 9–10.

9. (72) $x^2 + x - 12$

10. (75) $2x^2 + 8x + 8$

11. (45) Nancy is renting a kayak for 7 days. She can spend $125 at most. There is a service fee of $20. What can she spend at most on the per day rental fee including the $20 service fee?

Test 15–continued

12. (43) Determine the value for which the expression $\dfrac{(n-1)(n+5)}{3n-6}$ is undefined.

13. (67) Solve the system of equations below.

    $y = 2x - 3$

    $6x - 3y = -3$

14. (34) Determine whether the sequence below is an arithmetic sequence.

    −2, 1, 4, 7, . . .

    If yes, find the common difference and the next two terms.

15. (47) A toy store marks up the price of train sets they purchase at $25.00 each by 25%. What is the markup and new price of each train set?

16. (74) Solve the equation $|x| = 5$.

17. (60) A flag is in the shape of a rectangle. It has a length of $(y + 5)$ inches and a width of $(y - 5)$ inches. What is the area of the flag?

18. (57) Find the LCM of 6, 8, and 10.

Solve the inequalities in problems 19–20 and graph them on a number line.

19. (70) $-3x \geq 3$

20. (66) $x + 6 > 4$

Test 16

1. (67) Karen started walking at 3.5 miles per hour. After she walked 2 miles, her friend Ned started walking along the same route at a pace of 3.5 miles per hour. If they continue to walk at the same rate, will Ned ever catch up to Karen? Explain.

2. (71) State whether there is a positive correlation, a negative correlation, or no correlation between the data values in the scatter plot shown below.

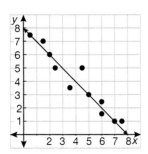

Simplify problems 3–4.

3. (69) $7\sqrt{xy} - 5\sqrt{xy}$

4. (76) $\sqrt{32}\sqrt{2}$

5. (60) Find the product $(a+4)(a-4)$.

6. (56) Write an equation for a direct variation that includes the point (3, 18).

7. (Inv. 8) Suppose y varies jointly with x and z. Find y when $x = 3$ and $z = 5$, given that $y = 72$ when $x = 6$ and $z = 4$.

Factor the trinomials in problems 8–10.

8. (79) $x^4 + 4x^3 + 3x^2$

9. (72) $x^2 + 8x + 15$

10. (75) $3x^2 - x - 2$

11. (70) A car salesperson earns a 12.5% commission on each car sold. What price must a car sell for in order for the salesperson to earn a commission of at least $2000?

Test 16–continued

12. (73) Write a compound inequality that describes the graph below.

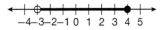

13. (78) Identify the asymptotes for the equation $y = \dfrac{2}{x-2}$.

14. (80) A student spins the spinner below and flips a fair coin. Make a table to show the possible outcomes in this experiment and find the theoretical probability of each outcome.

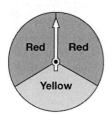

15. (68) Find the probability of rolling a sum of 3 or a sum of 11 on two number cubes, each numbered 1 to 6.

16. (74) Solve the equation $3|x| = 15$.

17. (52) Kaitlin is selling T-shirts. She sold 2 for $16 and 3 for $23. What will Kaitlin charge for 5 T-shirts if she keeps selling T-shirts at the same rate?

18. (42) What number is 140% of 35?

Solve the inequalities in problems 19–20 and graph them on a number line.

19. (77) $6x - 5 \leq -23$

20. (66) $n - 3 > -1$

Name _____   Date _____

Test 17

1. (84) A penny is dropped from a 512-foot tall tower. The equation $512 - h = 16t^2$ can be used to find the height h of the penny after falling for t seconds. Estimate the height of the penny after falling 2 seconds.

2. (85) Determine whether the side lengths below form a Pythagorean triple.

   9, 16, 25

Simplify problems 3–5.

3. (76) $\sqrt{3}\left(2 + \sqrt{6}\right)$

4. (58) $(2x + 4)(-3x^2 + 5x + 3)$

5. (61) $\sqrt{9s^4 t^5}$

6. (56) Tell whether the set of ordered pairs below represents a direct variation. If the set of ordered pairs represents a direct variation, find the constant of variation.
   (3, –15), (–2, 10), (–4, 20)

7. (65) Determine whether the lines are parallel.

   Line 1: $y = \dfrac{3}{5}x - 9$

   Line 2: $10x - 6y = 18$

Factor the trinomials in problems 8–10.

8. (79) $-2x^3 - 4x^2 + 16x$

9. (72) $x^2 + xy - 2y^2$

10. (75) $-22x + 12 + 8x^2$

11. (77) Jim borrows $95 from his parents. They agree to subtract $10 from the loan for each hour he does chores around the house. To find the number of hours x he needs to do chores before he owes less than $50, solve the inequality $95 - 10x < 50$.

Test 17–continued

12. (83) Determine whether the polynomial $x^2 + 18x + 81$ is a perfect-square trinomial. If it is, factor the trinomial.

13. (80) Suppose that 6 cards each contain one letter from the word BANANA. Make a bar graph to represent the frequency distribution for all possible outcomes. Find the theoretical probability of each outcome.

14. (78) Identify the asymptotes and graph the function below.

$$y = \frac{4}{x + 3}$$

15. (73) Solve the disjunction
$4x + 3 \leq 15$ OR $3x + 2 > -7$.

16. (69) Add $x\sqrt{3x} + \sqrt{27x^3}$.

17. (64) If y varies inversely as x and $y = 3$ when $x = 12$, find x when $y = 4$.

18. (53) Write the polynomial $4a^2 + 2a + a^3$ in standard form. Then write the leading coefficient.

Solve the inequalities in problems 19–20 and graph them on a number line.

19. (81) $4x + 10 \leq -2x - 8$

20. (82) $-8 \leq 4x - 2 + 2x \leq 10$

Name _____   Date _____

Test 18

1. (82) A farmer randomly chose 4 pumpkins out of a patch of 15 pumpkins. Three of the pumpkins weigh 15 lb, 18 lb, and 16 lb. What could be the weight of the fourth pumpkin if the average weight of all 4 pumpkins is to be between 15 and 20 lb?

2. (84) Use a table to graph the function $f(x) = 3x^2$.

3. (61) Simplify $\sqrt{1,000,000}$ using powers of ten.

4. (74) Solve $|x - 2| = 9$.

5. (76) Find the product $\left(5 - \sqrt{5}\right)^2$.

6. (69) A rectangular patio has a length of $\sqrt{363}$ feet and a width of $\sqrt{243}$ feet. What is the patio's perimeter?

7. (73) Write a compound inequality that represents all real numbers that are less than −1 or greater than 4. Graph the solution.

8. (68) Jed opens a drawer containing T-shirts. There are 5 white shirts, 3 blue shirts, 2 red shirts, and 2 gray shirts. If Jed picks a shirt at random, what is the probability he will pick a white or a blue T-shirt?

9. (86) Find the distance between the points (2, −3) and (5, 6).

10. (87) Factor $4x^2 + 8xy + 3x + 6y$.

11. (62) The stem-and-leaf plot below shows ages of students in a cooking class. Find the median, mode, range, and relative frequency of age 35.

Ages of Students in a Cooking Class

Stem	Leaves
1	6, 8, 8
2	4, 5, 5, 5, 6, 8
3	0, 1, 2, 5, 5, 5, 5, 8
4	0, 2, 9

Key 2 | 4 = 24

Test 18–continued

12. (83) Determine whether the binomial below is the difference of two squares. If it is, factor the binomial.

    $9x^2 - 100$

13. (85) Use the Pythagorean Theorem to determine the missing side length c.

    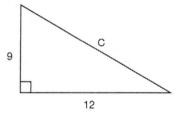

14. (Inv.9) Choose a factoring method for the polynomial $a^2 - 169$. Explain your choice. Then factor the polynomial.

15. (89) Give the coordinates of the parabola's vertex. Then give the maximum or minimum value and the domain and range of the function.

    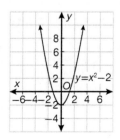

16. (88) Multiply $\dfrac{4a^3b^4}{3ab^2} \cdot \dfrac{9a^2b^2}{8a^2b}$.

17. (90) Add $\dfrac{y^2}{8y} + \dfrac{5y^2}{8y}$.

Solve the inequalities in problems 18–20 and graph them on a number line.

18. (77) $-4 + (-6) < -2y - 4$

19. (81) $4(x - 3) - 2x \leq 14 - 2(3x + 1)$

20. (70) $-3n \leq -9$

Name _____ Date _____

Test 19

1. (89) A ball is thrown from a height of 10 feet above the ground. The ball starts with a vertical speed of 64 feet per second. Ignoring friction, the equation $y = -16t^2 + 64t + 10$ gives the height y as a function of time t. Find the highest point the ball reaches and how long it takes to reach this point.

2. (84) Determine whether the equation below represents a quadratic function.

    $y + 7x^2 = -5x - 4$

3. (83) Determine whether the trinomial below is a perfect-square trinomial. If it is, factor the trinomial.

    $8x^2 - 8x + 2$

4. (86) Find the midpoint of the line segment with endpoints (−8, 5) and (2, 2).

5. (76) Find the product $\left(2 - \sqrt{7}\right)^2$.

6. (78) Identify the asymptotes for the rational function below.

    $y = \dfrac{3}{x + 7} + 2$

7. (73) Write a compound inequality that describes the graph below.

8. (68) What is the probability of rolling either a sum of 7 or a sum of 9 using two different number cubes, each numbered 1 to 6?

Factor the polynomials in problems 9–10.

9. (79) $3x^3 + 3x^2 - 18x$

10. (87) $2y^2 - 3y^3 - 9y + 6$

Test 19—continued

11. (92) It took Lee $\dfrac{2x^2 - 6x}{2x}$ minutes to drive to a mall that was $\dfrac{3x - 9}{x^3}$ miles away. Find her rate in miles per minute.

12. (95) Find the least common denominator (LCD) for the expression below.

$$\dfrac{2}{(x + 5)} - \dfrac{3}{(x^2 + 4x - 5)}$$

13. (85) A ramp extends from a 4-foot high platform to a point on the ground 20 feet away. Find the length r of the ramp in the diagram below. Round your answer to the nearest tenth of a foot.

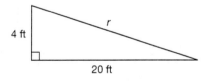

14. (94) Solve the equation below and graph the solution.

$$\dfrac{|x|}{6} + 4 = 16$$

Divide problems 15–16.

15. (93) $(8x^3 + 16x^2 + 6x) \div 2x$

16. (88) $\dfrac{5x^3y^3}{6xy^2} \div \dfrac{10x^2y^2}{9x^2y}$

17. (90) Add $\dfrac{3x^3}{2x^2} + \dfrac{2x}{4x}$. Simplify your answer.

Solve the inequalities in problems 18–20 and graph them on a number line.

18. (81) $\dfrac{3c}{4} + \dfrac{3}{8} \leq \dfrac{c}{4} - \dfrac{5}{8}$

19. (77) $-4(2 - x) \leq 2^2$

20. (91) $|x| < 8$

Test 20

1. (98) The area of a rectangular piece of plywood is 77 square feet. The length is 4 feet more than the width. What are the length and width of the plywood?

2. (96) Graph the function $y = x^2 + 4x + 2$.

Simplify problems 3–4.

3. (92) $\dfrac{\frac{m}{x}}{\frac{n}{m+x}}$

4. (76) $\sqrt{3}\left(\sqrt{6} - \sqrt{8}\right)$

5. (86) Find the distance between the points (6, −4) and (1, 1).

6. (84) Determine whether the graph of each function below opens upward or downward.

$f(x) = 4x^2$

$f(x) = -4x^2$

7. (100) Solve the equation below by graphing the related function.

$x^2 - 9 = 0$

8. (97) Determine whether the ordered pair (0, 3) is a solution of the inequality $y > 2x - 1$.

Factor the polynomials in problems 9–10.

9. (79) $-x^2 - x + 12$

10. (87) $20x^3y - 12x^3 + 10x^2y - 6x^2$

Test 20—continued

11. (99) It takes Ann 3 hours to rake a yard. It takes Linda 2 hours to rake a yard. How long will it take them if they work together?

12. (89) Find the axis of symmetry for the quadratic function $y = -2x^2 + 8x + 3$.

13. (85) Find side length t to the nearest tenth.

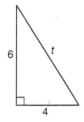

14. (94) Solve the equation.

$$\frac{3|x|}{6} + 4 = 2$$

15. (93) Divide $\left(x^2 - 10x + 25\right) \div \left(x - 5\right)$.

16. (88) Multiply $\dfrac{3a^2b + 2ab}{4a} \cdot \dfrac{14}{3ab + 2b}$.

17. (95) Add $\dfrac{3x^2}{x^2 - 9} + \dfrac{x - 2}{2x - 6}$.

Solve the inequalities in problems 18–20 and graph them on a number line.

18. (81) $3(x - 4) - x > 8 - 2(4x - 5)$

19. (82) $-4 < x + 5 + 2x \le 11$

20. (91) $\dfrac{|x|}{3} > 1$

Test 21

1. *(100)* Liz drops a marble from the top of a cliff 96 feet off the ground. The height of the marble is described by the quadratic equation $h = -16t^2 + 96$ where h is the height in feet and t is the time in seconds. Find the time t when the marble hits the ground. Round to the nearest hundredth.

2. *(96)* Graph the function $y = 2x^2 + 8x + 6$.

Simplify problems 3–4.

3. *(92)* $\dfrac{\frac{3}{x}}{1 + \frac{4}{x}}$

4. *(76)* $\left(9\sqrt{4}\right)^2$

5. *(93)* Charlie wants to find the length of a rectangular painting. The area is $\left(x^2 - 15x + 54\right)$ square inches. The width is $(x - 9)$ inches. What is the length of the painting?

6. *(104)* Complete the square.

   $x^2 + 4x + \underline{}$

7. *(103)* Rationalize the denominator of $\sqrt{\dfrac{5}{2}}$.

8. *(95)* Subtract $\dfrac{2x^2}{3x - 6} - \dfrac{4x - 3}{x^2 - 4}$.

9. *(79)* Factor $5x^3y + 25x^2y - 30xy$.

10. *(89)* Find the zeros of the function shown in the graph.

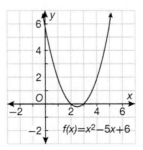

Test 21–continued

11. *(91)* A thermometer shows the temperature outside to be 67°F. The thermometer has an accuracy of plus or minus 2 degrees. Let the true temperature be *t*. Write an absolute value inequality to show that *t* is within 2 degrees of 67°F. Then solve the inequality. What is the range for the true temperature outside?

12. *(97)* Graph the inequality $y \geq -2x - 4$.

13. *(105)* Find the next four terms of the geometric sequence below.

    3, 6, 12, 24, . . .

Solve problems 14–17.

14. *(98)* $(x - 3)(x + 2) = 0$

15. *(99)* $\dfrac{4}{x} = \dfrac{2}{x - 2}$

16. *(102)* $x^2 = 169$

17. *(94)* $|4x| + 7 = 31$

18. *(86)* Find the midpoint of the line segment with endpoints (–8, 1) and (5, –9).

19. *(69)* Combine $\dfrac{3\sqrt{5p}}{11} + \dfrac{2\sqrt{5p}}{11} - \dfrac{6\sqrt{2q}}{11}$.

20. *(101)* Solve the inequality $4|x| + 2 < 10$ and graph it on a number line.

Test 22

1. (95) Kelly rides her bike at 14 mph if there is no wind. She plans a round trip to a park that is 35 miles away. If there is a w mph wind, the time for the outbound trip is $\dfrac{35}{14 + w}$ hours. The time for the return trip against a w mph wind is $\dfrac{35}{14 - w}$ hours. What is the total time for the round trip?

2. (107) Graph the function $f(x) = |x| + 3$ and give the coordinates of the vertex.

Simplify problems 3–4.

3. (103) $\dfrac{\sqrt{32x^4}}{2\sqrt{12x^3}}$

4. (92) $\dfrac{\dfrac{5x}{2x+6}}{\dfrac{10}{x+3}}$

5. (97) Determine whether the ordered pair $(-2, 10)$ is a solution of the inequality $y < -4x + 2$.

6. (90) Add $\dfrac{2x^2}{12x} + \dfrac{4x^2}{12x}$. Simplify your answer.

7. (Inv. 10) Write an equation for the transformation described below.

   Shift $f(x) = -2x^2 - 1$ up 3 units.

   Then graph the original function and the graph of the transformation on the same set of axes.

8. (93) Divide $(12x^3 + 24x^2 + 18x) \div 6x$.

9. (110) Use the quadratic formula to solve for x.

   $x^2 - 5x + 6$

10. (109) Graph the system of inequalities below.

    $y \geq \dfrac{1}{2}x + 2$

    $y \leq \dfrac{1}{2}x - 4$

Test 22–continued

11. (106) A garden has a vegetable section and a flower section as shown in the diagram below.

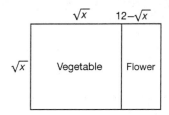

If the area of the garden is 108 square feet, what is the area of the flower section?

12. (104) Solve the equation $x^2 + 12x = 13$ by completing the square.

13. (105) The first term of a geometric sequence is 3 and the common ratio is −3. Find the 6th term of the sequence.

Solve problems 14–17.

14. (102) $x^2 + 10 = 19$

15. (94) $3\left|\dfrac{x}{2} - 5\right| = 12$

16. (99) $\dfrac{6}{x} = \dfrac{2}{x-6}$

17. (98) $(x-12)(x-9) = 0$

18. (108) Evaluate the function $f(x) = 4^x$ for $x = -1, 0,$ and 2.

Solve the inequalities in problems 19–20 and graph them on a number line.

19. (91) $|x| + 4 > 2$

20. (101) $|x + 4| - 3 > 2$

Test 23

1. *(111)* Celeste has four skirts: green, blue, purple, and red. Each skirt has 2 matching belts. Draw a tree diagram to determine the number of possible skirt and belt outfits that Celeste can wear.

2. *(114)* Determine the domain of $y = \sqrt{x - 12}$.

3. *(92)* Simplify the expression below.

$$\frac{\frac{xm}{ny}}{\frac{mn}{x}}$$

4. *(85)* Use the Pythagorean Theorem to find side length t to the nearest tenth.

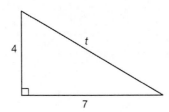

5. *(110)* Use the quadratic formula to solve for x.

$$2x + -24 + x^2 = 0$$

6. *(109)* Graph the system of linear inequalities below.

$$y \leq 3x + 2$$
$$y \geq \frac{1}{2}$$

7. *(107)* Describe the graph of the function $f(x) = 4|x|$.

8. *(93)* Divide $(x^2 - 7x + 10) \div (2 - x)$.

9. *(112)* Solve the system of equations below by substitution.

$$y = x^2 + 3x - 2$$
$$y = 3x + 7$$

10. *(115)* Evaluate the function $y = 3x^3$ for $x = -2, -1, 0, 1,$ and 2. Then graph the function.

Test 23—continued

11. *(Inv. 11)* Richard invested $1000 in a mutual fund that will double in value every 15 years. How many times will the amount double in 45 years? Write an equation to model the value of Richard's investment after x doubling times. What will be the value of the investment in 45 years?

12. *(95)* Find the least common denominator (LCD) for the expression
$$\frac{2}{(x-3)} - \frac{7}{(x^2+3x-18)}.$$

13. *(113)* Use the discriminant to find the number of real solutions to the equation $y = x^2 - 2x + 5$. Then state the number of x-intercepts of the graph of that equation $y = x^2 - 2x + 5$.

Solve problems 14–15.

14. *(94)* $\dfrac{5|x|}{2} - 6 = 4$

15. *(106)* $\sqrt{x} = 8$

16. *(102)* $x^2 = 400$

17. *(99)* $\dfrac{x}{6} = \dfrac{5}{x-1}$

18. *(108)* Determine whether the set of ordered pairs below satisfies an exponential function. Explain your answer.

    {(–1, 4), (1, 16), (–2, 2), (0, 8)}

Solve the inequalities in problems 19–20 and graph them on a number line.

19. *(91)* $|x| + 8 < 3$

20. *(101)* $|x+6| - 5 > 7$

Name _____ Test _____ Score _____

Test Answer Form A

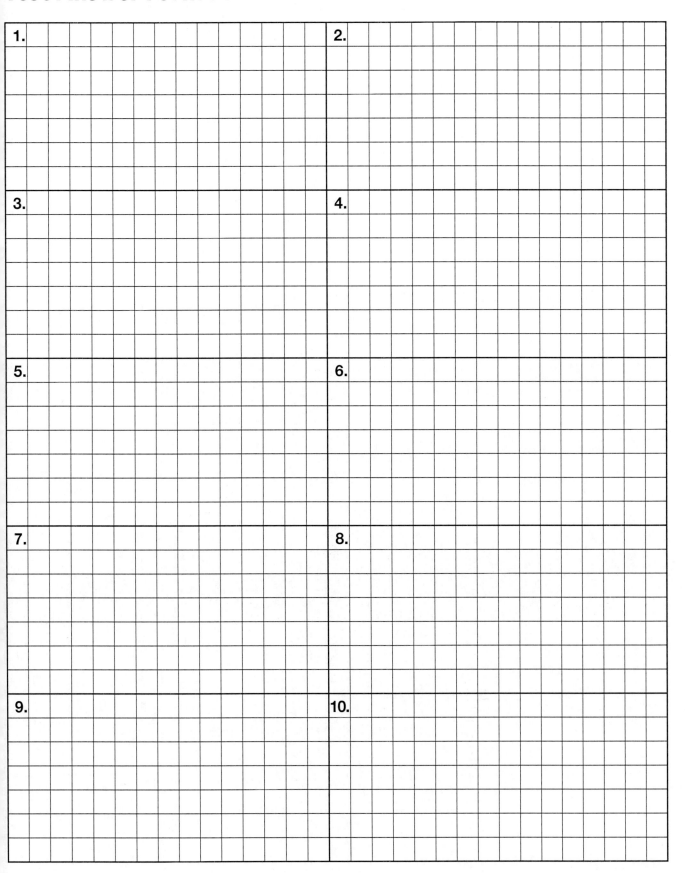

Name _____ Test _____ Score _____

Test Answer Form A–continued

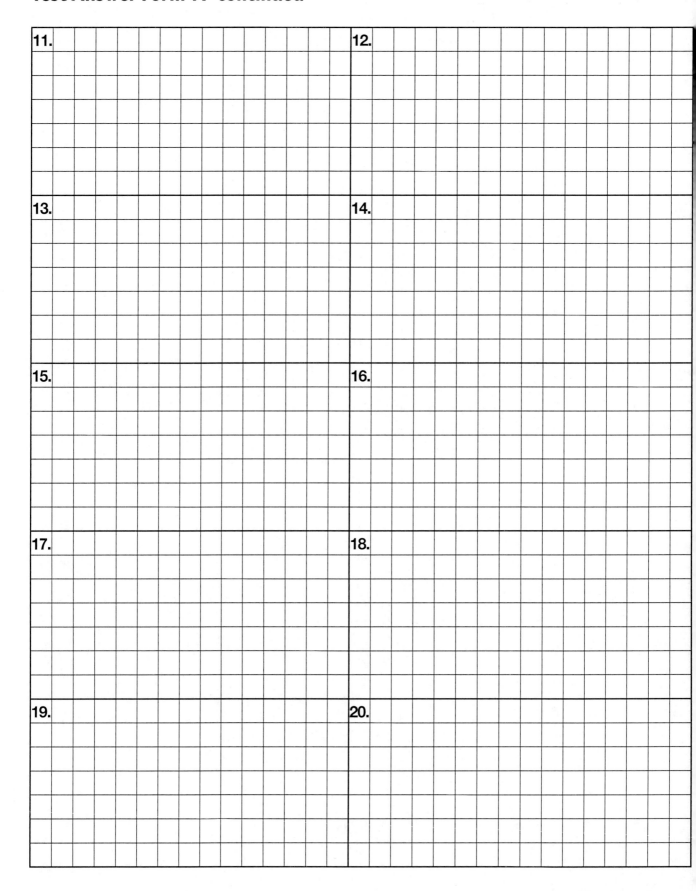

Name _____ Test _____ Score _____

Test Answer Form B

1.	2.
3.	4.
5.	6.
7.	8.
9.	10.

Name _____ Test _____ Score _____

Test Answer Form B—continued

11.	12.
13.	14.
15.	16.
17.	18.
19.	20.

Name _____   Test _____   Score _____

Test Answer Form C

1.	2.	3.	4.	1.	
				2.	
				3.	
				4.	
5.	6.	7.	8.	5.	
				6.	
				7.	
				8.	
9.	10.	11.	12.	9.	
				10.	
				11.	
				12.	
13.	14.	15.	16.	13.	
				14.	
				15.	
				16.	
17.	18	19.	20.	17.	
				18.	
				19.	
				20.	

Test Analysis Form

Test Item No.	Test Number											
	1	2	3	4	5	6	7	8	9	10	11	12
	Lesson Assessed											
1.	3	1	12	17	8	8	31	38	38	50	45	60
2.	2	4	9	16	2	22	23	Inv. 3	40	33	54	Inv. 6
3.	4	7	6	16	22	27	33	27	44	48	51	53
4.	2	2	Inv. 1	20	13	25	Inv. 2	37	45	2	48	58
5.	3	3	11	18	3	3	32	40	39	46	40	51
6.	4	4	11	18	4	4	10	39	32	39	18	46
7.	5	5	15	15	5	10	8	18	18	40	39	43
8.	3	3	15	11	7	5	12	11	12	38	32	39
9.	4	7	3	10	11	18	17	29	41	41	53	50
10.	3	1	10	7	15	29	30	36	42	44	44	57
11.	2	6	14	8	18	17	34	33	36	49	55	59
12.	1	6	13	19	25	14	20	8	43	47	49	44
13.	1	8	12	19	14	1	35	34	27	43	46	47
14.	1	2	Inv. 1	6	20	30	24	6	Inv. 4	37	31	45
15.	5	9	1	5	21	26	21	28	31	34	28	41
16.	5	9	5	13	23	28	28	26	28	28	23	42
17.	4	10	4	14	24	21	26	24	26	31	47	56
18.	2	10	8	17	19	23	13	10	23	Inv. 3	41	48
19.	5	8	14	Inv. 1	16	24	16	9	19	45	52	49
20.	1	5	13	20	9	16	29	17	35	Inv. 5	34	52

Test Analysis Form—continued

| Test Item No. | Test Number |||||||||||
	13	14	15	16	17	18	19	20	21	22	23
	Lesson Assessed										
1.	47	61	63	67	84	82	89	98	100	95	111
2.	61	31	71	71	85	84	84	96	96	107	114
3.	60	69	61	69	76	61	83	92	92	103	92
4.	58	7	40	76	58	74	86	76	76	92	85
5.	18	32	39	60	61	76	76	86	93	97	110
6.	49	58	46	56	56	69	78	84	104	90	109
7.	46	36	73	Inv. 8	65	73	73	100	103	Inv. 10	107
8.	45	65	65	79	79	68	68	97	95	93	93
9.	36	66	72	72	72	86	79	79	79	110	112
10.	65	70	75	75	75	87	87	87	89	109	115
11.	62	64	45	70	77	62	92	99	91	106	Inv. 11
12.	59	40	43	73	83	83	95	89	97	104	95
13.	63	63	67	78	80	85	85	85	105	105	113
14.	52	41	34	80	78	Inv. 9	94	94	98	102	94
15.	43	44	47	68	73	89	93	93	99	94	106
16.	57	47	74	74	69	88	88	88	102	99	102
17.	64	54	60	52	64	90	90	95	94	98	99
18.	48	68	57	42	53	77	81	81	86	108	108
19.	38	Inv. 7	70	77	81	81	77	82	69	91	91
20.	56	67	66	66	82	70	91	91	101	101	101

Test Solutions

Test 1

1. 10^{18} hertz

2. The factors are $\frac{1}{2}$ and m.
   The coefficient is $\frac{1}{2}$.

3. 57.76 square feet

4. The first term is $8jk$;
   the second term is $2x$;
   the third term is $\frac{4j}{3k}$.

5. $\frac{1}{121}$

6. 42

7. -2

8. $x^5 y^9$

9. 3

10. 0.125

11. There are two terms in the expression; the constants are 9.0 and 15.95; the variable is h.

12. {irrational numbers, real numbers}

13. The set of rational numbers best describes the situation since the amount may be a decimal amount.

14. The set of irrational numbers best describes the situation. Since the area of a circle is equal to the square of the radius multiplied by π, it will be an irrational number.

15. -2.2

16. $(-2) + 10 = 8$; 8°F

17. $(3.2 + 2) \div 4 + 3^2 < \frac{(9 + 23)}{2} - 8 \div 2$

18. The numbers 3 and 2 are constants because they never change. The letters a, b, and x are variables because they represent unknown numbers.

19. The statement is true because the sum of any two whole numbers will be a whole number.

20. False; counterexample: $3 \div 4 = 0.75$

Test 2

1. {natural numbers, whole numbers, integers, rational numbers, real numbers}

2. $3^2 - 2 \cdot 2.5 = (3 + 4) + 6 \div 3 - \frac{10}{2}$

3. $18 \div \left[(4-1)^2 - 3\right] > \left[8 - (5-3)^2 \cdot 2\right]$

4. Constants: 5 and 2; variables: x

5. 0.064

6. 27

7. 5

8. $x^9 y^6$

9. 24

10. Sample: The statement is false. A counterexample is $2 - 4$. The difference, -2, is not a natural number.

11. It will be at -131 feet in relation to the surface.

12.

13. 4 miles per hour

14. $\frac{16a}{(2+b)}$, c, $2ac$, and $\frac{b}{4}$

15. 27

16. 29

17. $-\frac{3}{4}$, -0.5, $\frac{1}{2}$, 0.7

18. Location A

19. 3 barrels

20. 10°F

Test 3

1. Commutative Property of Multiplication

2. $x^2 + y^3 < x^2y^2$

3. 1.2

4. $\frac{7}{20}$, 35%; $\frac{1}{4}$, 25%; $\frac{2}{5}$, 40%

5. 1.4

6. −9

7. 8

8. −16

9. $\frac{1}{49}$

10. $\frac{5}{13}$

11. $\frac{2}{5}$, 0.4, 40%

12. No, 48 is not a perfect square. There is no number that when multiplied by itself equals 48.

13. The statement is false. To illustrate the Identity Property of Addition, the statement should be $c + 0 = c$.

14. The event is as likely to happen as not to happen.

15. {irrational numbers, real numbers}

16. −19

17. 193.5 square inches

18. 4 miles per hour

19. {6}

20. The value of $\sqrt{17}$ to the nearest integer is 4.

Test 4

1. $c + 5$

2. −10

3. 8

4.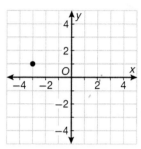

5. $3x$

6. $4n^2 + 5m^2$

7. $-8y + 56$

8. 24

9. −4.19

10. 139

11. 26 miles per hour

12. No, $x = 4$ is not a solution for the equation $x + 4 = 9$. The solution is $x = 5$.

13. $a = 15$

14. The set of integers is closed under subtraction.

15. −18

16. $\sqrt{16} + \sqrt{49} = \sqrt{25} + \sqrt{36}$

17. $\frac{2}{5}$, 0.4, or 40%

18. Sample: 9 less than m; the difference of m and 9

19. 0.250

20.

x	y
−2	0
0	4
2	8
$\frac{1}{2}$	5

Test 5

1. 70 miles per hour
2. $7ab$, $9c$, and $\frac{12a}{5b}$
3. median age is 38
4. side length is 15 feet
5. $x^{11}y^6$
6. 49
7. −8
8. 30
9. −0.5
10. $-x^3y + xy^2z^2$
11. $6a^4 + 5b^3$
12. $f(h) = 670 - 45h$
13. 0.8 or 80% or $\frac{4}{5}$
14. −8, 2, 12, 5
15. $x = -18$
16. $b = \frac{33}{10}$
17. $r = 5$
18. $y = -27$
19. 9
20. 88

Test 6

1. 54 minutes to weed the garden
2. Freshman: 126; Sophomore: 196; Junior: 140; Senior: 238
3. Sample: Because the scale does not start at zero, the number of customers appears to increase more rapidly than it actually did.
4. domain is {2, 6, 9}; range is {1, 4, 7, 8}
5. $\frac{1}{27}$
6. 27
7. $\frac{4}{9}$
8. $\frac{1}{3}$
9. $5a$
10. $y = -\frac{3}{5}x + \frac{2}{5}$
11. $\frac{4}{p} + 6$
12. {5, 6}
13. rational numbers, real numbers
14. domain is all real numbers and the range is $y \geq 2$
15. $a = 4$
16. $b = -1$
17. $y = -5$
18. $a = 3$
19. $s = 30$
20. −20

Test 7

1. $0.95 > $0.93; 5 lbs for $4.65 is the better buy.

2. 12 hours

3.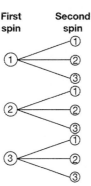

$P(2, 2) = \dfrac{1}{9}$

4. The graph is an upside-down U.

5. $\dfrac{1}{a^6 b^5}$

6. 10.15

7. Yes, 275 stones will be enough. Mr. Ramirez needs 270 stones to cover the patio.

8. Commutative Property of Multiplication

9. The quotient of 3 and p; 3 divided by p

10.

x	y
−2	−1
−1	0
0	1
1	2
2	3

The graph represents a linear function.

11. $c_n = 8 + (n - 1)5$; the total cost for 13 tickets is $68.

12.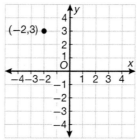

13. The x-intercept is 9. The y-intercept is 6.

14. 12.15

15. $x = 12$

16. $6c - 3 - 2c = 4c + 4$
$\quad\quad 4c - 3 = 4c + 4$ Combine like terms.
$\quad\quad\quad -4c = -4c$ Subtraction Property of Equality
$\quad\quad\quad\quad -3 = 4$

Since $-3 = +4$ is never true, the equation has no solutions.

17. $d = 4$

$5d + 2(2d + 3) = 42$
$\quad 5d + 4d + 6 = 42$ Distributive Property
$\quad\quad\quad 9d + 6 = 42$ Combine like terms.
$\quad\quad\quad\quad -6 = -6$ Subtraction Property of Equality
$\quad\quad 9d \cdot \dfrac{1}{9} = 36 \cdot \dfrac{1}{9}$ Multiplication Property of Equality
$\quad\quad\quad\quad\quad d = 4$ Simplify.

18. The value of $\sqrt{77}$ to the nearest integer is 9.

19. $\dfrac{3}{8}$

20. $s = 2t + \dfrac{5}{2}$

Test 8

1. $2 \cdot 5 \cdot 13$

2. Sample: Most members of the basketball team will name basketball as their favorite sport.

3. Sample: The large increments of the scale make the data values appear to be closer than they really are.

4. 6.48×10^7

5. $36x^4y^6$

6. $\dfrac{n^3x^2y}{m} + \dfrac{3}{mn^3x^4}$; $m \neq 0, n \neq 0, x \neq 0$

7. $3a^2 + ab^2$

8. -16

9. $15°C$

10. 18 feet

11. $\dfrac{1}{15}$

12. 1127 kilometers

13. Yes, arithmetic sequence; common difference is 11; 45, 56

14. 4.5

15. The equation is an identity. It has infinitely many solutions.

16. $b = -3$

17. $c = -7$

18. $-2, -\dfrac{1}{2}, -0.2, 0.2$

19. 63

20. $3n$

Test 9

1. $4x^2y^4z^2$

2. $121n^2$ square feet

3. $-\dfrac{3}{2}$

4. Sample: $3n > 5$

5. $\dfrac{2a^3}{b^2} + \dfrac{ab^2}{c}$; $b \neq 0, c \neq 0$

6. y^3

7. $3xy^2 + 2x^2y$

8. $9x$

9. 3 miles per hour

10. $937.50

11. $x = 2.5$ feet

12. $a = -5$

13. The large increments make the data values appear to be closer than they really are.

14. A student likes football, but he does not play on the team.

15. $y = 13$

16. $n = -1$

17. $z = 8$

18. $a = -\dfrac{2}{3}$

19. $a = \dfrac{13}{20}$

20. The x-intercept is 60. The y-intercept is 40. The x-intercept shows that if no games were rented, 60 movies were rented. The y-intercept shows that if no movies were rented, 40 games were rented.

Test 10

1. [number line with point at 0, marks from -4 to 4]

2. $\dfrac{1}{5}$

3. mean is 43; median is 42; mode is 44

4. factors are $\dfrac{1}{7}$ and x; coefficient is $\dfrac{1}{7}$

5. $\dfrac{4}{9}$

6. $\dfrac{x^2y^2z}{ab} + 3x^2y - \dfrac{b^2xy}{a}$

7. $16a^8b^4$

8. $7x^2y^4(x-y)$

9. 2

10. $-\dfrac{1}{4}$

11.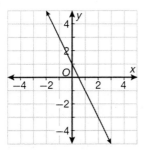

12. markup is $22.50; new price is $52.50

13. The expression is undefined for $a = -3$

14. 5.6×10^{-3}

15. -4, 1, 6, and 11

16. The health club costs are the same for 20 months.

17. 14 men; 10 women

18. A possible bias is that people who enter the library but do not check out a book will not be surveyed.

19. Sample: Six is less than or equal to the product of 9 and b.

20. Sample: If a figure has four sides, then it is a quadrilateral.

Test 11

1. $18

2. No heights are less than 45 inches or greater than 85 inches, so there are no outliers.

3. $\dfrac{x}{4}$; $x \neq 0$

4. The weight of 3 pounds is an outlier. Without that weight, the mean is about 39 pounds and the median is 39 pounds. With that weight, the mean is about 34 pounds and the median is 38 pounds.

5. $8x^9y^3$

6. $-1x^3 + 3y^2 + 2x^3y^2$

7. $mn^3 + 4n$

8. c^2

9. $2x^3 + 4x^2$; leading coefficient is 2

10. $-\dfrac{1}{5}$

11. The ordered pair only makes $x - 2y = 6$ true. (2, 4) is not a solution of the system.

12. $y = \dfrac{1}{3}x + 2$

13. 2 inches

14. $z = 15$

15. The equation has infinitely many solutions.

16. $t = -6$

17. markup is $5.40; new price is $17.40

18. The rate of change is 2 pages per minute.

19. $y - 1 = 2(x - 3)$

20. Yes, arithmetic sequence; common difference is -5; next two terms are -14 and -19

Test 12

1. $x^2 - 9$ square inches

2.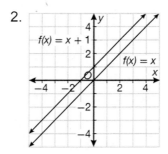

The graph of $f(x) = x + 1$ shifts up 1 unit from the graph of $f(x) = x$. The y-intercept shifts from (0, 0) to (0, 1).

3. 5

4. $-6a^3b - 12ab^2 - 9abc$

5. $\dfrac{-x}{d^3}$

6. 5

7. $\dfrac{y}{3}$, $y \neq 4$

8. $\dfrac{c^3}{a} + ab^2c$

9. $\{-1, 0, 1, \text{and } 2\}$

10. LCM is 280

11. (1, 3)

12. $-\dfrac{3}{4}$

13. discount is $23.75; new price is $71.25

14. sample: $\dfrac{5}{8}n \leq 15$

15. The slope is 0.

16. 99

17.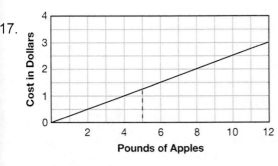

The cost of 5 pounds of apples is about $1.25.

18. The range for Class A is 42. The range for Class B is 48. Class B has a greater range of scores.

19. slope is $-\dfrac{3}{4}$; y-intercept is 3

20. $14

Test 13

1. discount is $27; new price is $63

2. $3\sqrt{2}$

3. $y^2 - 49$

4. $8xy^3 - 6x^2y - 10xyz$

5. $3m^2n^2 - 2m^2n$

6.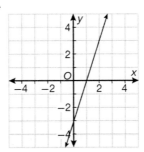

7. 6 inches

8. $12{,}306 \leq 3000n$

9. 60°

10. $y = 2x + 7$

11. Key 4|8 = 48 inches

Heights of Children at Summer Camp (in inches)

Stem	Leaf
3	6, 8, 9, 9
4	2, 4, 5, 5, 5, 7, 8, 8, 8, 9
5	0, 2, 2, 3, 4, 4, 6, 9
6	0, 0, 0, 1, 2, 2, 2, 5

12. (−2, 4)

13. (3, 4)

14. $y + 3 = 3(x - 3)$

15. $x = 2$

16. 24 days

17. $x = 14$

18. The score of 67 is an outlier. With the outlier, the mean score is about 20 and the median is 14. Without the outlier, the mean score is about 14 and the median is 13.5. The mode of 17 stays the same.

19. $-4x^2y^3(y + x)$

20. $y = 4x$

Test 14

1. $2\sqrt{7}$ inches

2. $\dfrac{1}{2}$ mile per minute

3. $-7\sqrt{xy}$

4. -157

5. $\dfrac{1}{a^2b^4}$

6. $-6x^2 + 12x - 6$

7. 30 feet

8. Yes, the lines are parallel since they both have a slope of $\dfrac{2}{5}$.

9. $x < 2$

10. $x > -6$

11. 6 hours

12. $25n^2$ square centimeters

13. $(2, 1)$

14. 35 miles per hour

15. -3

16. discount is $8.75; new price is $26.25

17.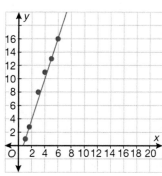

Half of the students scored between 65 and 85.

18. $\dfrac{5}{18}$

19. $k = 2;\ y = 2x$

20. Since the graphs are the same line, there are infinitely many solutions.

Test 15

1. 9 plums

2.

$y = 3x - 2$

3. $9cd\sqrt{d}$

4. $9x^6y^4$

5. $n + 5mn$

6. $\dfrac{2}{11}$

7. $x > 1$ AND $x < 6$ or $1 < x < 6$

8. Yes, the lines are perpendicular since their slopes are negative reciprocals of each other. Line 1 has a slope of $\dfrac{3}{2}$ and line 2 has a slope of $-\dfrac{2}{3}$.

9. $(x+4)(x-3)$

10. $2(x+2)^2$

11. $15

12. 2

13. The system has no solution. It is an inconsistent system.

14. Yes; common difference is 3; 10, 13

15. markup is $6.25; new price is $31.25

16. $\{5, -5\}$

17. $y^2 - 25$ square inches.

18. 120

19. $x \leq -1$

20. $x > -2$

Test 16

1. No, Ned will not catch up to Karen. The situation can be represented by the system of equations below:

   $y = 3.5x$ and $y = 3.5x + 2$; the lines are parallel, so there is no solution.

2. negative correlation

3. $2\sqrt{xy}$

4. 8

5. $a^2 - 16$

6. $y = 6x$

7. $y = 45$

8. $x^2(x+3)(x+1)$

9. $(x+3)(x+5)$

10. $(3x+2)(x-1)$

11. $16,000

12. $x > -3$ AND $x \leq 4$ or $-3 < x \leq 4$

13. Vertical asymptote: $x = 2$; horizontal asymptote: $y = 0$

14.

	Red	Yellow	Red
Heads	RH	YH	RH
Tails	RT	YT	RT

$P(R, H) = \frac{1}{3}$; $P(R, T) = \frac{1}{3}$; $P(Y, H) = \frac{1}{6}$;

$P(Y, T) = \frac{1}{6}$;

15. $\frac{1}{9}$

16. $\{5, -5\}$

17. $37

18. 49

19. $x \leq -3$

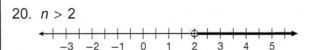

20. $n > 2$

Test 17

1. 448 feet

2. No, $9^2 + 16^2 \neq 25^2$

3. $2\sqrt{3} + 3\sqrt{2}$

4. $-6x^3 - 2x^2 + 26x + 12$

5. $3s^2 t^2 \sqrt{t}$

6. Yes, the constant of variation is -5.

7. No, the lines are not parallel because the slope of line 1 is $\frac{3}{5}$ and the slope of line 2 is $\frac{5}{3}$.

8. $-2x(x-2)(x+4)$

9. $(x-y)(x+2y)$

10. $2(4x-3)(x-2)$

11. Jim needs to do chores for more than $4\frac{1}{2}$ hours.

12. Yes, it is a perfect-square trinomial; $(x+9)^2$

13.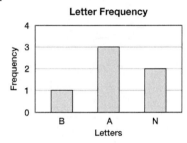

$P(B) = \frac{1}{6}$; $P(A) = \frac{1}{2}$; $P(N) = \frac{1}{3}$

14. Vertical asymptote: $x = -2$; horizontal asymptote: $y = 0$

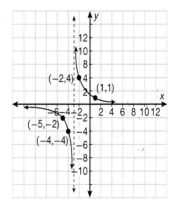

15. $-3 < x \leq 3$

16. $4x\sqrt{3x}$

17. $x = 9$

18. $a^3 + 4a^2 + 2a$; the leading coefficient is 1.

19. $x \leq -3$
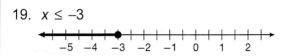

20. $-1 \leq x \leq 2$

Test 18

1. between 11 lb and 31 lb

2.

x	−2	−1	0	1	2
f(x)	12	3	0	3	12

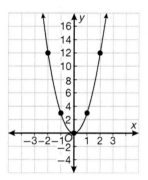

3. 1000

4. {−7, 11}

5. $30 - 10\sqrt{5}$

6. $40\sqrt{3}$ feet

7. $x < -1$ OR $x > 4$
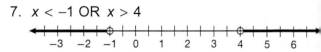

8. $\frac{2}{3}$

9. $3\sqrt{10}$

10. $(x+2y)(4x+3)$

11. median: 30.5; mode: 35; range: 33 years; relative frequency of 35: 20%

12. Yes, the binomial is the difference of two squares; $(3x+10)(3x-10)$

13. $c = 15$

14. Difference of two squares; there are only two terms, they are being subtracted, and they are perfect squares; $(a+13)(a-13)$

15. vertex: $(0, -2)$; minimum: -2; domain: set of all real numbers; range: all real numbers greater than or equal to -2

16. $\dfrac{3a^2b^3}{2}$

17. $\dfrac{3y}{4}$

18. $y < 3$

19. $x \leq 3$

20. $n \geq 3$

Test 19

1. 74 feet; 2 seconds

2. yes

3. yes; $2(2x-1)^2$

4. $(-3, \dfrac{7}{2})$

5. $11 - 4\sqrt{7}$

6. vertical: $x = -7$; horizontal: $y = 2$

7. $-2 < x \leq 4$ or $x > -2$ AND $x \leq 4$

8. $\dfrac{5}{18}$

9. $3x(x+3)(x-2)$

10. $(y^2 + 3)(2 - 3y)$

11. $\dfrac{3}{x^3}$

12. $(x+5)(x-1)$

13. 20.4 feet

14. $x = -72$ or $x = 72$

15. $4x^2 + 8x + 3$

16. $\dfrac{3x^2}{4}$

17. $\dfrac{3x+1}{2}$

18. $c \leq -2$

19. $x \leq 3$

20. $-8 < x < 8$

Cumulative Test 20

1. length: 11 feet; width: 7 feet

2.

3. $\dfrac{m(m+x)}{xn}$

4. $3\sqrt{2} - 2\sqrt{6}$

5. $5\sqrt{2}$

6. $f(x) = 4x^2$ opens upward; $f(x) = -4x^2$ opens downward

7.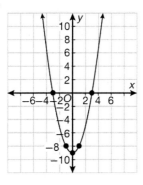

   The solutions are 3 and −3.

8. yes

9. $-(x + 4)(x - 3)$

10. $2x^2[(5y - 3)(2x + 1)]$

11. $1\frac{1}{5}$ hours

12. $x = 2$

13. 7.2

14. no solution; or $\varnothing$

15. $x - 5$

16. $\frac{7}{2}$

17. $\dfrac{7x^2 + x - 6}{2(x - 3)(x + 3)}$

18. $x > 3$

19. $-3 < x \leq 2$

20. $x > 3$ OR $x < -3$

Test 21

1. $t = 2.45$

2.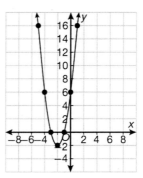

3. $\dfrac{3}{x + 4}$

4. 324

5. $(x - 6)$ inches

6. $x^2 + 4x + 4$

7. $\dfrac{\sqrt{10}}{2}$

8. $\dfrac{2x^3 + 4x^2 - 12x + 9}{3(x - 2)(x + 2)}$

9. $5xy(x + 6)(x - 1)$

10. The zeros are 2 and 3.

11. $|t - 67| \leq 2$; $65 \leq t \leq 69$; the range is between 65°F and 69°F.

12.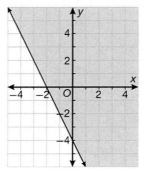

13. 48, 96, 192, 384

14. $\{-2, 3\}$

15. $x = 4$

16. $x = \pm 13$

17. $\{6, -6\}$

18. $(-\dfrac{3}{2}, -4)$

19. $\dfrac{5\sqrt{5p} - 6\sqrt{2q}}{11}$

20. $-2 < x < 2$

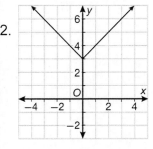

Test 22

1. $\dfrac{980}{(14 + w)(14 - w)}$ hours

2.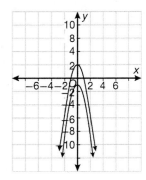

    The vertex is (0, 3).

3. $\dfrac{\sqrt{6x}}{3}$

4. $\dfrac{x}{4}$

5. No, 10 is not less than 10.

6. $\dfrac{x}{2}$

7. $f(x) = -2x^2 + 2$

    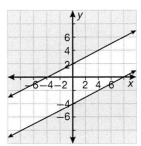

8. $2x^2 + 4x + 3$

9. $x = 2$ and 3

10.

    The two solution sets do not intersect so the system has no solution.

11. 27 square feet

12. $x = 1$ or $x = -13$

13. -729

14. $x = \pm 3$

15. $\{2, 18\}$

16. $x = 9$

17. $\{9, 12\}$

18. $f(-1) = \dfrac{1}{4}$; $f(0) = 1$; $f(2) = 16$

19. The set of all real numbers $\mathbb{R}$;

20. $x < -9$ OR $x > 1$

Test 23

1.
Skirt	Belt	Outcomes
Green	1	Green Belt 1
	2	Green Belt 2
Blue	1	Blue Belt 1
	2	Blue Belt 2
Purple	1	Purple Belt 1
	2	Purple Belt 2
Red	1	Red Belt 1
	2	Red Belt 2

8 possible outcomes

2. $x \geq 12$; the set of all real numbers greater than or equal to 12

3. $\dfrac{x^2}{n^2 y}$

4. 8.1

5. $x = 4$ or $x = -6$

6.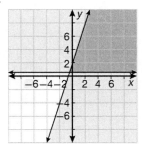

7. $a = 4$, so $|a| = 4$; since $|a| > 1$, the graph is stretched vertically. Since $a > 0$ the graph opens upwards.

8. $-x + 5$

9. (3, 16) and (−3, −2).

10.
x	−2	−1	0	1	2
y	−24	−3	0	3	24

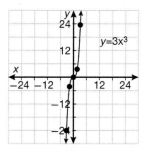

11. 3; $y = 1000(2)^x$; $8000

12. $(x + 6)(x - 3)$

13. No real solutions; no x-intercepts

14. $\{-4, 4\}$

15. $x = 64$

16. $x = \pm 20$

17. $x = 6$ or $x = -5$

18. Yes; As the x-values increase by the constant amount 1 the ratio of the y-values is 2; the base $b = 2$.

19. $\{\ \}$ or $\varnothing$

20. $x < -18$ OR $x > 6$